I Ching (Yi Jing) and Modern Science

I Ching (Yi Jing) and Modern Science

Its Application for the Benefit of Human Society

CHARLES H CHEN, M.D., FACOG

I CHING (YI JING) AND MODERN SCIENCE
ITS APPLICATION FOR THE BENEFIT OF HUMAN SOCIETY

iUniverse books may be ordered through booksellers or by contacting:

iUniverse
1663 Liberty Drive
Bloomington, IN 47403
www.iuniverse.com
1-800-Authors (1-800-288-4677)

ISBN: 978-1-5320-2870-0 (sc)
ISBN: 978-1-5320-2871-7 (e)

Print information available on the last page.

iUniverse rev. date: 09/07/2017

Contents

Abstract

Yi Jing is created from keen observation of the nature in the universe. The key components of Yi Jing are the yin and yang duality, the four phenomena, trigram, two sets of eight trigrams, the 64 hexagrams, and two sets of 64 trigrams.

The yin and yang duality of the universe posited in Yi Jing correlates with the matter duality of the universe theorized in quantum mechanics.

The four phenomena are two sets of four yin and yang configurations, they correlate with two sets of four chemical configurations found in DNA and RNA.

Trigram is a triple yin and yang configuration, it correlates with codon which is three-chemical configuration in DNA and RNA.

Two sets of 64 trigrams in Yi Jing correlate with two sets of 64 codons in DNA and RNA. Yi Jing's 64 hexagrams consist of two sets of 64 trigrams which are the 64 primary trigrams and the 64 secondary trigrams, they correlate with the two sets of 64 codons which are the DNA 64 codons and the RNA 64 codons.

Yi Jing's external factors of changes (64 hexagrams and 384 lines) are the key principles that offer the seekers advice accurately and wisely for dilemmatic issues in oracle use.

Introduction

Chinese Emperor Fu Xi (2852–2737 BC) created the *I Ching/Yi Jing* [1](Wade-Giles pronunciation/Chinese Pinyin pronunciation) about 5,000 years ago. King Wen of Zhou (1099–1050 BC) and his son, the Duke of Zhou (1050–1046 BC), completed the final version of the text about 3,000 years ago. Darwin's theory of evolution :changing factors[2] was proposed in the nineteenth century, while DNA/RNA :genetic code[3] and quantum mechanics :matter duality[4] were discovered in the twentieth century, placing these advances some 3,000 years after the creation of Yi Jing.

The book describes the eight primary trigrams created by Fu Xi[5], which refer to the existing things in the universe, and the secondary trigrams discovered by King Wen of Zhou[6], which refer to the changes that interact and interrelate with the existing things in the universe.

Yi Jing is the only ancient book that is grounded in

---

[1] Richard Wilhelm, *The I Ching or Book of Changes*, trans.(Princeton, NJ: Princeton University Press, 1967)

[2] Charles Darwin, *On the Origin of Species by Means of Natural Selection* (London, England: John Murray, 1859)

[3] Geoffrey M.Cooper and Robert E.Hausman, *The Cell—A Molecular Approach* (Sunderland, MA:Sinauer Associates, 5th Edition, 2009)

[4] Walter Greiner, *Quantum Mechanics: An Introduction* (Springer; 2001) Note: matter of universe in this context means the material of existing things in the universe.

[5] Richard Wilhelm, *The I Ching or Book of Changes*, trans.(Princeton, NJ: Princeton University Press, 1967), pp. 265-67

[6] Richard Wilhelm, *The I Ching or Book of Changes*, trans.(Princeton, NJ: Princeton University Press, 1967), pp. 268-72

observer-based science. It is consistent with modern scientific findings related to the universe, including those of object-based science (DNA and RNA) and quantum mechanics (Matter Duality). Compiled over a period of 2,000 years, from the days of Emperor Fu Xi to those of King Wen of Zhou, Yi Jing is considered one of the richest treasures of Chinese wisdom.

Worldwide, scholars of Yi Jing agree that it is one of the most important books in world literature; however, these scholars have been unable to explain how the key components of Yi Jing –- including the yin and yang duality, the four phenomena, trigrams, two sets of eight trigrams, the 64 hexagrams, and two sets of 64 trigram — are consistent with modern scientific findings. This is because few scholars possess knowledge of both Yi Jing and broad modern science. This author, however, acquires a broad understanding of modern science and will, through an in-depth analysis of Yi Jing, explain in this paper how Yi Jing is consistent with modern scientific findings.

In Yi Jing, 'Yi' means changes and, in the book, change conveys the following three meanings.

- Everything is in the process of continuous change.[7]
- The underlying causes of change concern many external factors of changes / lines that produce different outcomes.[8]

---

[7] Hellmut Wilhelm, *Understanding The I Ching : The Wilhelm Lectures on the Book of Changes* (Princeton, NJ: Princeton University Press, 1988), pp. 236-7

[8] Richard Wilhelm, *I Ching or Book of Changes*, trans.(Princeton, NJ: Princeton University Press, 1967), pp. 312-3

- The 64 hexagrams are arranged in various sequences of cyclic change, in line with the nature of the universe.[9]There are 384 lines in 64 hexagrams.

Following his constant keen observation of the nature in the universe, Emperor Fu Xi perceived the idea of the Taiji [10] as oneness. He expressed the idea of the Taiji as a simple, solid line that broke into two lines. He designated the solid line as yang and the broken line as yin. The yin and yang duality[11] correlate with the matter duality[12] in quantum mechanics. Emperor Fu Xi perceived yin and yang as having two aspects/sides[13] paired together, which produce four phenomena; this is calculated by $2 \times 2 = 4$.[14] The four phenomena correlate with the four chemicals in DNA and RNA. Fu Xi observed many varieties of things on earth and therefore believed that another element would change the balance of yin and yang in the universe. A triple yin and yang configuration would produce many varieties of things in the universe. With this in mind, he would add

---

[9] Alfred Huang, *The Complete I Ching* (Rochester, VT: Inner Tradition, 2010), pp. xxi

[10] John Blofeld, *I Ching: The Book of Change* (London, England: Penguin Group, 1965), p. 49

[11] Richard Wilhelm, *The I Ching or Book of Changes*, trans.(Princeton, NJ: Princeton University Press, 1967), p. lv

[12] Walter Greiner, *Quantum Mechanics: An Introduction.* (Springer: 4th Edition, 2001)

[13] Richard Wilhelm, *I Ching or Book of Changes*, trans.(Princeton, NJ: Princeton University Press, 1967), p. lv

[14] Richard Wilhelm, *I Ching or Book of Changes,* trans.(Princeton, NJ: Princeton University Press, 1967), p. 319.Note: The translation of Chinese "Hsiang" in this context should be Phenomenon, not image.

either another yin or another yang to the yin and yang duality in order to change the balance. He then created the trigram, which is a triple yin and yang configuration. A trigram correlates with a codon in DNA and RNA. By taking three lines out of two to form a trigram, the maximum number of permutations is eight ; this is calculated by 2 x 2 x 2 = 8.

Fu Xi created eight primary trigrams, each comprising three yin and yang configurations. He also perceived that the universe consisted of eight fundamental elements: heaven, thunder, water, mountain, earth, wind, fire, and lake.[15] He matched each of the eight trigrams to each of these fundamental elements.

During the Chinese Zhou dynasty, some 3,000 years ago, King Wen of Zhou[16] proposed that external factors of change constantly interact with everything that exists in the universe, changing the fate of those things. With this in mind, he added eight secondary trigrams to complement Fu Xi's eight primary trigrams. King Wen of Zhou thus formulated 64 hexagrams ; this is calculated by 8 x 8 = 64. These 64 hexagrams consisted of two sets of 64 trigrams which correlate with the DNA and RNA 64 codons[17].

These 64 hexagrams were composed of two sets of 64 trigrams. The two sets of 64 trigrams have not appeared in Chinese history books; they derived from mathematical

---

[15] Richard Wilhelm, *The I Ching or Book of Changes*, trans.(Princeton, NJ: Princeton University Press, 1967), p. 265

[16] Richard Wilhelm, *The I Ching or Book of Changes*, trans.(Princeton, NJ: Princeton University Press, 1967), pp. 268-9

[17] Geoffrey M.Cooper and Robert E.Hausman, The Cell—A Molecular Approach. (Sunderland, MA: Sinauer Associates, 5th Edition, 2009)

configurations ; they are deduced by this author. The first set comprised Fu Xi's eight primary trigrams; this is calculated by 4 x 4 x 4 = 64. This permutation was achieved by taking three lines out of four to form a trigram, with the four lines deriving from the four phenomena. The first 64 trigrams correlate with the 64 codons in DNA.

The second set of 64 trigrams was King Wen of Zhou's eight secondary trigrams; this is calculated by 4 x 4 x 4 = 64. This permutation was achieved by taking three lines out of four to form a trigram, with the four lines again deriving from the four phenomena. The second 64 trigrams correlate with 64 codons in RNA.

In the twenty-first century, scientists established the theory of observer-based science, grounded in the idea that the observer is crucial in understanding the universe. This theory is similar to the theories of quantum mechanics[18], which involves the study of physical phenomena at the microscopic scale.

During the nineteenth and twentieth centuries, scientists sought to describe the make-up of living things. Charles Darwin (1809–1882) discovered the external factors of change that altered and changed the fate of living things. The Yi Jing, however, had already deduced what living things were made of, as well as the external factors of change that influenced and affected their fate. With the advancement of perspective science, the paradigm shifted from observer-based science to object-based science and quantum mechanics.

---

[18] Walter Greiner, *Quantum Mechanics: An Introduction.* (Springer; 2001).

Human understanding regarding the universe progressed from the observable universe represented by Fu Xi and Wen of Zhou, who created the Yi Jing, to the molecular universe represented by James Watson and Francis Crick, who formulated the structure of DNA and RNA, to the invisible universe represented by German scientist Max Planck, who pioneered quantum theory. Each applied different scientific approaches in order to answer the same questions about the universe.

Yi Jing

Emperor Fu Xi created Yi Jing about 5,000 years ago. The final version of Yi Jing was completed during the Zhou dynasty by King Wen of Zhou and his son the Duke of Zhou; the text also became known as Zhou Yi.

The key components of Yi Jing are the yin and yang duality, the four phenomena (two sets of four yin and yang configurations), trigram, eight primary trigrams and eight secondary trigrams, the 64 hexagrams, and two sets of 64 trigrams.

This author acquires a broad understanding of modern scientific theories relating to the universe, and is well-suited to describing how, 5,000 and 3,000 years ago, respectively, Emperor Fu Xi and King Wen of Zhou postulated the key components of *Yi Jing* that is consistent with modern scientific findings regarding the universe.

Creation of Yin and Yang

When Fu Xi reigned over the dynasty in China, Chinese society had not yet developed written communication. Fu Xi taught many things in Chinese society, such as how to fish with nets and hunt with iron weapons. He knew that true wisdom comes from direct and keen observation of the nature in the universe. He gazed upwards and contemplated the heavens, and then looked downwards and contemplated what happened on earth. Many people asked him for guidance to avoid natural disasters or environmental changes that could determine their fate.

Fu Xi perceived the idea of the Taiji [19] as oneness. As there was no written language at that time, Fu Xi expressed the idea of the Taiji as a simple, solid line which then broke into two lines. He designated the solid line as yang and the broken line as yin. He perceived the notion of yin and yang duality as a symbolic fundamental property of the universe.[20]

---

[19] John Blofeld, *I Ching: The Book of Change.*(London, England: Penguin Group, 1965), p. 49

[20] John Blofeld, *I Ching: The book of Change.*(London, England: Penguin Group, 1965), p. 49

Creation of the Four Phenomena

Following the creation of yin and yang, Fu Xi posited that each yin and yang had two aspects/sides[21], which paired with one another, producing four phenomena; this is calculated by 2 x 2 = 4. These are old yang, young yang, old yin, and young yin.[22] The old yang is composed of two solid lines. The young yang is composed of one solid line at the bottom and one broken line at the top. The old yin is composed of two broken lines. The young yin is composed of one broken line at the bottom and one solid line at the top. The four phenomena were two sets of four yin and yang configurations.

---

[21] Richard Wilhelm, *The I Ching or Book of Changes*, trans.(Princeton, NJ: Princeton University Press, 1967), p. lv
[22] Richard Wilhelm, *The I Ching or Book of Changes*, trans.(Princeton, NJ: Princeton University Press, 1967), p. 319 Note: The translation of Chinese 'Hsiang' in this context should be phenomenon, not image.

Trigram Creation

Fu Xi observed the heavens above, the earth below, and everything in between. There were many different morphologies and functions for all living things on earth, and these intrigued Fu Xi. He thought that a separate entity must be responsible for changing the balance of yin and yang in the universe and producing the innumerable living things that exist in the world. Based on this idea, he developed a three-line configuration. He used solid line as yang and broken line as yin. He added either a yin or a yang to change the balance of yin and yang. Thereafter, he created three yin and yang configurations, such as yang-yin-yang or yin-yang-yang. This was the so-called trigram.[23]

---

[23] Hellmut Wilhelm, *Understanding The I Ching: The Wilhelm Lectures on The Book of Changes* (Princeton, NJ: Princeton University Press, 1988), p. 49 "Dao(Tao)gave birth to the one, the one gave birth to the two, the two gave birth to the three, and the three gave birth to all things" from Dao De Jing by Laozi. His Aphorisms were inspired by Yi Jing.(Dao De Jing or Tao Te Ching)

Creation of Eight Primary Trigrams

By taking three lines out of two to form a trigram, such as yang-yin-yang or yang-yang-yin, the maximum number of permutations was eight; this is calculated by 2 x 2 x 2 = 8. These were the eight primary trigrams, with the three-line configuration, created by Fu Xi.

Creation of Eight Primary Trigrams with Eight Trigram Names

With his constant observation of the universe, Fu Xi perceived that there were eight fundamental elements: heaven /qian, thunder /zhen, water /kan, mountain /gen, earth /kun, wind /xun, fire /li, and lake /dui. He assigned the names of each of the eight fundamental elements of the universe to eight trigrams according to the symbol/image of each trigram. He then paired those eight elements.

The first pairs were heaven, with three solid lines, and earth, with three broken lines. The second pairs were lake, with two solid lines at the bottom and one broken line at the top, and mountain, with two broken lines at the bottom and one solid line at the top. The third pair was fire, with one broken line between two solid lines, and water, with one solid line between two broken lines. The fourth pair was thunder, with one solid line at the bottom and two broken lines at the top, and wind, with one broken line at the bottom and two solid lines at the top.

About 2,000 years after the time of Fu Xi, King Wen of Zhou assigned ancient Chinese ideograms to the eight trigrams: qian, zhen, kan, gen, kun, xun, li and dui. Those names depicted the functional aspects of the eight fundamental elements.[24]

Why did King Wen of Zhou use the functional aspects

---

[24] Richard Wilhelm, *The I Ching or Book of Changes*, trans.(Princeton, NJ:Princeton University Press, 1967), p. 1

of natural objects/fundamental elements to name the eight trigrams? The natural objects were fixed, and had unchanging properties. The functional aspects of the natural objects were not fixed, and had properties that could change with time. This is similar to quantum mechanics[25], which describes physical systems as being divided into types according to unchanging properties, in this instance, heaven or earth; and a system at one time consisted of a complete specification of those of its properties that changed with time, in this instance, qian or kun.

---

[25] Stanford Encyclopedia of Philosophy, *Quantum Mechanics* (First published in 2000; Substantive revision in 2015), p. 2

Yi Jing's 64-Hexagram Formulations

About 3,000 years ago (2,000 years after Fu Xi), during the Shang Dynasty (1766–1121 BC), a written form of Chinese emerged. King Wen of Zhou[26] meditated on Fu Xi's eight primary trigrams and began observing the nature of the universe. He perceived that the external factors of change[27] were interacting with existing things in the universe brought on by the passing of time, and were thereby changing or altering their fates.

This observation intrigued King Wen of Zhou; he then added eight secondary trigrams to complement Fu Xi's eight primary trigrams. The eight primary trigrams represented the existing things in the universe; the eight secondary trigrams represented the external factors of change interacting with the existing things in the universe. The maximum number of permutations when combining two sets of eight trigrams is 64; this is calculated by 8 x 8 = 64. These were the 64 hexagrams formulated by King Wen of Zhou. Each had six lines, and each set of three lines represented one image. Hence, two images were combined. For example, hexagram seven has an image of water at the bottom and one of earth at the top.

---

[26] Richard Wilhelm, *The I Ching or Book of Changes*, trans.(Princeton, NJ: Princeton University Press, 1967), pp. 268-9.The secondary eight trigrams show the progression of change in the phenomenal world, and they are the patterns of environmental changes.

[27] Richard Wilhelm, *The I Ching or Book of Changes*, trans.(Princeton, NJ: Princeton University Press, 1967), pp. 268-72

There were two sets of 64 trigrams. These were derived from mathematical configurations; they are deduced by this author, they do not show up in the records of Chinese history books. The first set was Fu Xi's eight primary trigrams; this is calculated by 4 x 4 x 4 = 64. The second set was King Wen of Zhou's eight secondary trigrams; this is calculated by 4 x 4 x 4 = 64. By taking three lines out of four to form a trigram, the maximum number of permutations was 64; this is calculated by 4 x 4 x 4 = 64. The four lines were derived from the four phenomena.

Yi Jing's 64 Hexagrams (Names and Sequence with the Number Formulation)

First, the 64 hexagrams were arranged in order from 1 to 64. Each of the 64 hexagrams was composed of six lines: the lower three from Fu Xi's trigrams, and the upper three from King Wen of Zhou's trigrams. The first hexagram was composed of six solid lines /yang. In the second hexagram, one broken line /yin replaced the bottom line. For the third hexagram, the broken line moved to the second place from the bottom. After the broken line reached the top line at the end of such a cycle, there are total 6 lines, the bottom line then was replaced with two broken lines until it reached the top line at the end of a second cycle. Subsequently, the cycles ended at number 64, which was composed of six broken lines.[28]

King Wen of Zhou assigned a name to each hexagram using the ancient Chinese writing system of ideographs. Those 64 hexagram names were assigned according to the mutual relationship between the upper and lower trigrams. For example, the name of the first hexagram was qian which is two heavens and full of power. The name of the second hexagram was gou which is heaven above and wind below, they are coming together or encountering. After names were assigned to all 64 hexagrams, Wen of Zhou rearranged them in sequential order, which is the order

---

[28] Richard Wilhelm, *The I Ching or Book of Changes*, trans.(Princeton, NJ: Princeton University Press, 1967), pp. 730-31

in which they can be found on the current 64-hexagram table. This was done systemically, according to the cyclic changes at various sequences in line with the nature of the universe.[29] The Chinese ideographical names are ordered according to the meaning and position of that ideograph in the cycle.

After Wen of Zhou named the 64 hexagrams, he numbered them. The first hexagram was qian, which referred to the initiation or creation of the universe. The second hexagram was kun, which referred to responding or receptive. The qian and kun, first and second hexagrams, were the basis from which the other 62 hexagrams were named and numbered.

---

[29] Alfred Huang, The Complete I Ching (Rochester, VT: Inner Tradition, 2010), p. xxi

Formulation of the 64-Hexagram Judgement /Decision Texts

King Wen of Zhou wrote the judgement /decision texts that were appended to each hexagram. Those texts contained his interpretation of the meaning of the hexagrams. The ideograms of the hexagrams' names represented each hexagram's main theme.

Yi Jing's 384-Line Text Formulation

Following King Wen of Zhou's 64-hexagram formulation, his son, the Duke of Zhou, wrote interpretative texts for each line of the hexagrams.

There were 64 hexagrams, and each hexagram had six lines. Combined, the 64 hexagrams comprised 384 lines; this is calculated by 64 x 6 = 384. The six lines were commensurate with hexagram's main theme, and each contributed a particular meaning and guidance to the main hexagram. When Yi Jing in use as an oracle, a seeker casted a coin or yarrow stalk, there were changing lines[30] involved during casting; therefore, the actual number of changing factors was more than 384. A hexagram could also change into another hexagram through the changing lines. The 64 hexagrams and 384 lines encompassed all situations that pertain the various complex conditions in human society on earth.

Throughout history, many Chinese sages have written commentaries on Yi Jing. One was the famous Chinese sage Confucius, who wrote the Ten Wings[31]. These Ten Wings provide a useful interpretation of the contents and contexts of Yi Jing.

---

[30] Neyma Jahan, *The Celestial Dragon I Ching* (London, England: Watkins Publishing, 2012), p. 17 Note: changing lines here mean when seeker throws coin, it shows all three coins are whether head or tail. This is a changing line.

[31] Richard Wilhelm, *The I Ching or Book of Changes*, trans.(Princeton, NJ:Princeton University Press, 1967), pp. 256-60

Confucius once said, "If some years were added to my life, I would give fifty to the study of the book of Yi Jing and might then escape falling into great errors."[32]

Even he was overwhelmed by Yi Jing. This shows how complicated the Yi Jing texts are.

---

[32] James Legge, *I Ching : Book of Changes* (New York, NY: Gramercy Books, 1996), p. xi

DNA and RNA

The Swiss physiological chemist Dr Friedrich Miescher discovered DNA in the late 1860s. The American James Watson and Englishman Francis Crick[33] formulated an accurate description in 1953, and received the Nobel Prize for physiology/medicine in 1962.

A cell contains both DNA and RNA. DNA is the combination of four chemicals: adenine (A), thymine (T), guanine (G), and cytosine (C). RNA is the same, except uracil (U) replaces DNA's thymine (T). Those two sets of four chemicals are the fundamental building blocks for every living thing on earth.

DNA is a tightly coiled, two-stranded helix; the two strands are connected by hydrogen bonds. RNA is single-stranded. These two sets of four chemicals always come in pairs, such as A-T or C-G, called base pairs. The DNA inside a cell consists of many millions or billions of bases. A segment of DNA is known as encoded DNA, or a gene.

A gene is a unit of inheritance that carries information. Genes are specific sequences of bases that encode instructions for making a body's proteins. Proteins are essential components of life.

---

[33] James Watson and Francis Crick, *A Structure for Deoxyribose Nuclei Acid.* (Nature 171, 1953)

How DNA and RNA Process, Transform, and Develop into Life

Life consists of morphologies and functions. DNA and RNA make proteins for living things, including human beings. Such proteins are responsible for perpetuating life, functions, physical developments, and behaviors.

A gene is a segment of DNA coding for a particular protein of a living cell.[34]

- A gene makes up a coding sequence called exon, and a non-coding sequence called intron. Some genes contain constitutive promoters and always turn on. Some genes contain regulated promoters. These regulated promoter genes determine when and how genes will turn on and begin the processes to make proteins. Those regulated promoter genes become active in response to specific stimuli. Those stimuli come from external factors of change, such as nutritional environments, or hot or cold weather environments; they can change the morphologies and functions of living things. These regulated promoter genes play a major role in cellular evolution.
- Transcription is the first step in transforming DNA and RNA from lifeless chemicals into a living entity.

---

[34] Geoffrey Cooper and Robert Hausman, *The Cell: A Molecular Approach* (Sinauer Associates, Inc.; 5th Edition, 2009) Note: The paragraph here is the summary of a cell's protein synthesis described by this author.

An enzyme called RNA polymerase unwinds the section of DNA that carries the desired gene. Only one strand of DNA is recognized. RNA polymerase binds to a promoter site and begins to form an RNA copy of the template strand. This RNA is called messenger RNA or mRNA.

- The mRNA consists of a nucleotide triplet, called a codon, such as A-U-G or A-C-G. A nucleotide is composed of one chemical base attached to a sugar and a phosphate.

- Each codon consists of three chemicals. DNA or RNA is a combination of four chemicals. By taking three chemicals out of four to form a codon, the maximum number of permutations is 64; this is calculated by $4 \times 4 \times 4 = 64$. Therefore, DNA has 64 codons, and RNA also has 64 codons. Genetic code consists of these two sets of 64 codons.

- The completed single strand of mRNA migrates out of a cell's nucleus through pores in the nuclear membrane. Then it binds to the ribosome in the cell's cytoplasm. This RNA is called ribosome RNA or rRNA.

- Translation is the second and final step in transforming lifeless chemicals into a living being. Each mRNA or codon codes for a particular amino acid. Amino acids are brought to the ribosome in the cytoplasm of a cell by transfer RNA or tRNA and linked in the correct sequence to the mRNA according to the triplet code. Amino acids then join together to form a polypeptide chain which is the primary structure of a protein. The order in which the amino acids join together is determined by the original gene.

During the process of amino acid synthesis, all but two of the amino acids — methionine and tryptophan — can be encoded by two to six different codons. A codon is a basic operative unit that processes and produces amino acids. Proteins are assembled from amino acids. All living things are made out of proteins.

There are 20 standard amino acids in the human body. Nine essential amino acids come from the foods we eat, and the body can produce the other 11 non-essential amino acids; these can be encoded from 64 different codons. Therefore, those two sets of 64 codons— the so-called genetic code— are the essential chemical structures that produce all living things on earth.

Every sequence of DNA and RNA codons can be read in three reading frames. Each produces a different amino acid sequence. There are six reading frames in DNA: three in the forward orientation on one strand, and three in the reverse orientation on the opposite strand.

Darwinian Evolution

How Darwinian Evolution Works

Charles Darwin proposed the theory of evolution, which states that all life (including humans) evolved into its present form through the process of natural selection / external changing factors.

Each cell within a living thing is composed of a nucleus, cytoplasm and a membrane, and contains DNA and RNA chemical structures. DNA and RNA is the combination of four chemicals that form a building base for a cell. The DNA inside a cell consist of million or even billion bases. There is a segment of DNA known as encoded DNA or a gene. A gene gives instructions on how to make a cell's proteins.

When changing factors such as sunlight, weather, or nutritional environments come into contact with a cell, they will interact with the cell's DNA and RNA chemical structures. The external factors such as these can alter the DNA and RNA sequence of chemical structures. When the DNA and RNA sequence of a chemical structure changes, its cells will produce different proteins. Under these conditions, the morphology and function of a living thing can change.

Living things can change their morphologies and functions as a result of external stimuli. They do this in order to adapt to their environments. This is the idea upon which Darwin's theory of evolution is based. This process of changing the morphologies and the functions of living things may take many years to accomplish, but it depends on the external stimuli. Short-term events can also change the fate of a living being.

Darwin's theories regarding cellular evolution, following the formation of cellular life on our planet, reflex external stimuli found on earth. The mechanism of Darwin's cellular evolution by natural selection[35] is due to the following three processes and outcomes:

1. The gene alterations of the cells have a detrimental effect on the organisms. Those organisms will die out and disappear from our planet.
2. The gene alterations of the cells have a beneficial effect on the organisms. Those organisms will survive and change their functions and morphologies to be better adapted to change in their environment. This is what Darwin's theory of evolution is based upon.
3. The gene alterations of cells have a neutral effect on the organisms. Those organisms do not change their functions or morphologies.

The organisms will go through all three different processes again and again. The constant flux in environmental factors on our planet influences and changes living things. The organisms will survive up to the present time in the second and third processes.

---

[35] Ronald A.Fisher, *The Genetical Theory of Natural Selection* (Oxford, England: The Clarendon Press, 1930) Note: The paragraph here is the summary of cellular evolution due to genetic alteration described by this author.

Principle Architectures of Yi Jing and DNA/RNA

Yi Jing's trigram uses a triple yin and yang configuration, such as yang-yin-yang or yin-yin-yang. A codon in DNA or RNA is a nucleotide triplet, or a three-chemical configuration, such as A-U-G or A-C-G. The trigram in Yi Jing correlates with the codon in DNA or RNA, which is the basic operative unit that produces all living things.

The four phenomena in Yi Jing consist of two sets of four yin and yang configurations. These are old yang, young yang, old yin, and young yin. DNA is a combination of four chemicals: adenine (A), thymine (T), guanine (G), and cytosine (C). RNA is a combination of four chemicals: adenine (A), uracil (U), guanine (G), and cytosine (C). DNA and RNA consists of two sets of four chemical configurations.

Yi Jing's four phenomena correlate with the four chemicals in DNA and RNA. Yi Jing's four phenomena always come in pairs, such as yang-yin or yin-yang, as do the four chemicals in DNA and RNA, such as A-T or G-C.

Genetic Code

In Yi Jing, there are two sets of 64 trigrams. The first set is Fu Xi's 64 primary trigrams. The second set is King Wen of Zhou's 64 secondary trigrams.

In DNA and RNA, there are two sets of 64 codons. The first are those in DNA; the second are those in RNA. Fu Xi's 64 trigrams correlate with DNA's 64 codons. King Wen of Zhou's 64 trigrams correlate with RNA's 64 codons.

Fu Xi's 64 trigrams depict the existing things on earth. DNA's 64 codons are the genetic codes that produce all existing living things on earth.

King Wen of Zhou's 64 trigrams depict the external factors affecting and altering the fate of living things on earth. RNA's 64 codons are the 64 mRNA structures. External factors can change the chemical structures of mRNA. This will change the protein production of cells. Subsequently, the morphologies, functions, and ultimately the fate of living things change.

Principle Architectures of Yi Jing and Darwin's Theory of Evolution

Yi Jing

Yi Jing's external factors of changes interact constantly with living things brought on by the passing of time. Therefore, Yi Jing's external factors of changes produce different fates for living things.

Darwin's Theory of Evolution

Evolution has changed the morphology and function of living things since cellular life was formed some 3.8 billion years ago.

External factors of changes as interpreted in Yi Jing, or by Darwin's theory of evolution, whether they manifest as short term events or long term processes, ultimately change the fates of living things on earth.

Principle Architectures of Yi Jing (Yin and Yang Duality) and Quantum Mechanics (Matter Duality)

Yi Jing :Yin and Yang Duality

Yi Jing depicts yin and yang duality as the symbolic fundamental property of the universe. All yang has a yin nature and all yin has a yang nature; they complement each other.[36]

Quantum Mechanics :Matter Duality

Quantum mechanics depicts matter duality as the fundamental property of the universe. All particles have a wave nature and all waves have a particle nature; they complement each other.[37]

---

[36] Alfred Huang, *The Complete I Ching* (Rochester, VT: Inner Tradition, 2010), p. 4, 54

[37] Walter Greiner, *Quantum Mechanics: An Introduction* (Springer; 2001)

The Application of Yi Jing for the Benefit of Human Society

Yi Jing is considered to be the accurate oracle[38] and, as demonstrated by Dr. Carl G.Jung[39] that it can offer important insights to the questions of dilemmatic issues. This renders the advice it contains as objective, which means that it provides neutral, independent analyses of dilemmatic issues.

There are two types of oracle. The first type is the ancient oracle. These were people or agents who offered advice or prophecies thought to have come directly from a divine being or God, using a type of divination or fortune-telling. Those people believed that miracles must, therefore, come from a divine being or God. With respect to Yi Jing, the terms 'divination' and 'fortune-telling' have been used mistakenly in ancient times, and even in present times.

The second type of oracle is any good source of information or wisdom. Yi Jing is used as this type of oracle. The term I used as the wisdom oracle. Yi Jing's 64 hexagrams and 384 lines are the changing factors that can achieve accurate advice to the seekers; they have not come from a divine being or God.

---

[38] Hellmut Wilhelm, *Understanding The I Ching: The Wilhelm Lectures on The Book pf Changes* (Princeton, NJ: Princeton University Press, 1988), p. 120
[39] Richard Wilhelm, *The I Ching or Book of Changes*, trans.(Princeton, NJ: Princeton University Press, 1967), pp. xxvi—xxix

Yi Jing in Use as The Book of Oracles and Wisdom

Yi Jing in use as the book of oracles[40]; oracles are used to gain advice and guidance regarding questionable, difficult or dilemmatic issues. Human wisdom and knowledge is limited, it is impossible for anyone to know everything during his or her lifetime. Yi Jing composes many changing factors, this will enable people to choose and act accordingly. It is therefore useful and very accurate in use as oracle; however, if people use for other purposes, such as fortune-telling, it lacks moral significance.[41] Yi Jing is also not appropriate for divination, as it contains scientific findings and many changing factors, and is not derived from a divine or godly source.

Yi Jing in use as the book of wisdom[42]; the philosophy of Yi Jing contains the three important concepts; the first is the idea of change, second is its theory of ideas, and the third is the judgements. Two branches of Chinese philosophy, Confucianism and Laozi (author of Tao Te Ching), have their common roots from Yi Jing's wisdom. Yin and Yang duality lies at the origins of many branches of classical Chinese science and philosophy. The trigram in Yi Jing is the fundamental unit that produce all things in the universe.

---

[40] Richard Wilhelm, *The I Ching or Book of Changes*, trans.(Princeton, NJ: Princeton University Press, 1967), p.xlix

[41] Richard Wilhelm, *The I Ching or Book of Changes*, trans.(Princeton, NJ: Princeton University Press, 1967), p. liii

[42] Richard Wilhelm, *The I Ching or Book of Change*, trans.(Princeton, NJ: Princeton University Press, 1967), pp. liv—lviii

It is consistent with the nucleotide triplet(codon) in DNA and RNA. The codon(three A, T, G, C, U configuration) is the fundamental unit that produce all things in the universe. It also is consistent with the triad (three notes) in music world. The triad is the fundamental unit that can create many varieties of music songs. Yi Jing composes many changing factors (64 hexagrams and 384 lines), this will enable seekers to act wisely. Why the Yi Jing inventor can have such a great knowledge; I realize such a great knowledge must derive from their keen and intense observation of the nature in the universe and they were born with excellent brain cell.

Who Benefits from Yi Jing in Use as an Oracle

Anyone who is faced with a dilemma or difficult issue, including national leaders, institutional leaders, corporate executives, officials, and other individuals, can benefit from Yi Jing.

The followings explain the reasons why Yi Jing's advice to the seekers with accuracy:

1. Yi Jing's wisdom is grounded in observer-based science. It is consistent with modern scientific findings regarding the universe. It is not derived from a divine source or God. It is not a mystic, mythic, superstitious, or fortune-telling book.

2. Yi Jing offers objective, neutral, independent analyses for seekers. An analysis of the six lines will yield one of three outcomes: in general, the first two lines suggest unfavorable conditions, the second two lines suggest neutral conditions, and the third two lines suggest favorable conditions. If the seeker casts and gets a particular hexagram, and that hexagram is not applicable to his or her present situation, then the seeker does not need to take that advice.

3. Yi Jing contains 384 lines, which address states of change. The six lines of each hexagram pertain to various complex conditions in society, and contribute to the guidance of each main hexagram. Because there are changing lines during casting, there are more than 384 factors of change to cover all sorts of

situations that pertain to various complex conditions in human society.

Yi Jing's oracle system is similar to the ideal political system in the US, whose constitutional policies are analogous to Yi Jing's 64 hexagrams. The American president is analogous to a seeker; the main issue for discussion is analogous to the main hexagram; and the members of Congress are analogous to the lines.

The president will bring the main issue for discussion, and the members of Congress will contribute their opinions. However, the president has to make the final decision on the main issue.

How Yi Jing Works as an Oracle

All living things on earth are composed of two properties. The first is constitutive, which exists in every living thing from the moment of its creation. In Yi Jing's 64 hexagrams, this is under the descriptions of Fu Xi's eight primary trigrams. The second is the state-dependent property, which can change with time. In Yi Jing's 64 hexagrams, this is under the descriptions of King Wen of Zhou's eight secondary trigrams. The first property of living things cannot change; however, the second property of living things can change with time.

It is important to understand the physiology of human beings following their appearance on earth. Human beings have two components pertaining to the ways in which they deal with various complex situations and conditions in human society. These consist of conscious and unconscious processes. The conscious process enables a human being to create or learn from an experience. The unconscious process has a universal quality; it was also called the 'collective unconscious' by the famous Swiss psychiatrist Dr Carl G.Jung [43]; it is the universal datum, and is innate. One cannot acquire this stratum through education or conscious effort. Collective unconscious is common to all human beings' experience, regardless of time, space, age, and different race. We may describe it as

---

[43] Carl G.Jung, *The Archetype and The Collective Unconscious* (Princeton, NJ: Princeton University Press, 1981)

a universal library of human knowledge. Yi Jing consists of symbols/images that stand for more than their obvious or immediate meaning, they hint at something not yet known. Symbols come from a common source, that is the "collective unconscious"[44]. King Wen of Zhou and his son, the Duke of Zhou, interpreted these symbols/images with texts that expressed the collective unconscious of humanity. The texts are complex and obscure, and Chinese sages and scholars have struggled to interpret them for the past thousand years or so. This is because the symbols/images were applied to the universal quality of collective unconscious, which is open to interpretation. This is what makes the texts of the hexagrams and lines relevant to humanity today, and why Yi Jing still can use as an oracle and give seekers with accurate and wise advice.

When Yi Jing is used as an oracle, external factors of change become important. The changes have no consciousness, no action, they are quiescent and do not move. But when seekers cast the three coins or 50 yarrow stalks to get a hexagram, they can get the answers from the result of those they cast.[45] The six lines address the states of change that are commensurate with the main hexagram. It is important to know that individual hexagrams do not have a fixed or invariable meaning; their significance will

---

[44] Mark O'Connell Raje Airey and Richard Craze. *The Illustrated Encyclopedia of Symbols, Signs, and Dream Interpretation* (New York, NY: Annes Publishing, Metro Books, 2013), p. 251

[45] Richard Wilhelm, *The I Ching or Book of Changes*, trans.(Princeton, NJ: Princeton University Press, 1967), p. 315

become apparent only under the circumstances in which seekers cast.

The 64 hexagrams and 384 lines encompass all the situations that pertain to various complex conditions in human society. Author want to emphasize, seekers use Yi Jing when they are looking for advice only in the dilemmatic issue, but not the straightforward issue or fortune-telling issue. Author emphasize in this research papers that Yi Jing is like the encyclopedia of wisdom. In order to get answers from the 64 hexagrams and 384 lines, one must cast. If the seeker only reads Yi Jing, then Yi Jing cannot provide advice.

Every occurrence of casting a hexagram is linked to the moment in time in which it is done, as well as the dilemma facing the seeker at that time. This is termed "Synchronicity" by Carl G. Jung.[46] Synchronicity takes the coincidence of events in space and time as meaning something more than mere chance. Each hexagram is commensurate with six lines. The six lines address the states of change; each line contributes to the particular meaning and guidance of the main hexagram. If there is a changing line when seeker casts, then it will change the original hexagram and line to a new hexagram and line. The system therefore covers various complex conditions. One must consider all of the information and advice provided in each hexagram and its six lines alongside one's individual situation. Then one can decide whether to accept the advice.

---

[46] Richard Wilhelm, *The I Ching or Book of Changes*, trans.(Princeton, NJ: Princeton University Press, 1967), p. xxiv

Using Yi Jing for oracle is analogous to practicing medicine. In the treatment of a complex and difficult disease, a doctor gives a person several treatment options from which to choose. It is up to that person to make the final decision. But that person has to visit the doctor first. That person cannot read a medical textbook to find the treatments on his or her own.

There are not any oracle system in the whole world that have such a wisdom/scientific method to give seekers advice accurately and wisely.

Methods of Consulting the Oracle

There are many methods of consulting the oracle, but two are practiced most commonly. Before one casts a hexagram, they must prepare the questions carefully. With regard to the wording of the questions, the caster must make sure Yi Jing is treated as an advisor, but not as a fortune-teller to avoid making errors. It is also important to remember that the goal of casting a hexagram is to get an answer for a dilemmatic issue. If the issue is straightforward, with no factors of change involved, Yi Jing should not be consulted.

- The yarrow-stalk oracle has been practiced since around 500 BC. It is a complicated and cumbersome method, involving a process of selecting and sorting 50 stalks. Four stages (one-set operation) are required. These four stages are repeated three times, in order to form a line. A three-set operation will create one line. In order to produce six lines for one hexagram, six three-set operations are required.
- The three-coins oracle is simpler, and has been practiced since the Tang Dynasty (618–907 AD) or Sung Dynasty (1127–1279 AD). Three coins are tossed six times to form six lines. The first toss is for the bottom line. The second toss is for the second-to-bottom line. This occurs line by line until six lines are created. This will produce one of 64 hexagrams.

Conversion from Tossing Coins to Form a Line

In Yi Jing, six is the symbol of old yin, eight is the symbol of young yin, seven is the symbol of young yang, and nine is the symbol of old yang. Seven and eight are stable lines. Those lines are disregarded in interpreting the oracle. The negative line that moves by the number six and the positive line that moves by the number nine must be used when interpreting the oracle. The head side of the coin is yang and has a value of three. The tail side of the coin is yin and has a value of two.

One head and two tails form a solid line /yang[47]. Two heads and one tail form a broken line /yin. Three heads or three tails yield a changing line. Three heads change a line from a solid line /yang to a broken line /yin. Three tails change a line from a broken line /yin to a solid line /yang. If there is a changing line, one must read the interpretation on the original line and hexagram as well as a new changing line and a new hexagram.

After a hexagram is created, that hexagram's text, judgement, image, lines, and commentaries are looked up to arrive at a decision. Each individual's situation is different, and each individual has to make his or her own decision regarding whether to act after casting the hexagram and getting an answer. Yi Jing cannot make the decision for you.

After forming a hexagram, the individual can interpret the

---

[47] Neyma Jahan, *The Celestial Dragon I Ching* (London, England: Watkins Publishing, 2012), p.14

text themselves or consult a Yi Jing expert. The hexagrams and lines can be somewhat difficult to understand; even for Yi Jing experts, it takes many years to understand them.

The table of Yi Jing's 64 hexagrams, texts, judgements, images, lines, and commentaries is readily available worldwide.

Bibliography

Blofeld, John.*I Ching or The Book of Change* (London, England: Penguin Group, 1965)

Cooper, Geoffrey M, and Robert E hausman. *The Cell—A Molecular Approach* (Sunderland, MA: Sinauer Associates, 2013)

Darwin, Charles.*On the Origin of Species; evolution by natural selection* (London, England: John Murray, 1859)

Fisher, Ronald.*The Genetical Theory of Nature Selection* (Oxford, England: the Clarendon Press, 1930)

Greiner, Walter. *Quantum Mechanics:An Introduction* (Springer, 2001)

Huang, Alfred. *The Complete I Ching* (Rochester, Vermont: Inner Tradition, 2010)

Jahan, Neyma.*The Celestial Dragon I Ching* (London, England: Watkins Publishing, 2012)

Jung, Carl.*The Archetype and Collective Unconscious* (Princeton, New Jersey: Princeton University Press, 1981)

Legge, James.*I Ching; Book of Changes* (New York, NY: Gramercy Books, A Division of Random House, Inc. 1996)

O'Connell, Mark, Raje Airey and Richard Craze.*The Illustrated Encyclopedia of Symbols, Signs, and Dream Interpretation* (New York, NY: Annes Publishing, Metro Books, An Imprint of Sterling Publishing, 2013)

Stanford Encyclopedia of Philosophy, *Quantum Mechanics* (First Published in 2000; substantive revision in 2015)

Watson, James and Francis Crick.*A Structure for Deoxyribose Nuclei Acid* (Nature 171, 1953)

Wilhelm, Hellmut.*Understanding The I Ching; The Wilhelm Lectures on The Book of Changes* (Princeton, New Jersey: Princeton University Press, 1988)

Wilhelm, Richard.*The I Ching or Book of Changes* (Princeton, New Jersey: Princeton University Press, 1967), translated(from German to English) by Cary F. Baynes.

Printed in the United States
By Bookmasters